Instagram:

Everything You Need to Know About Instagram for Business and Personal Use - Ultimate Instagram Marketing Book

J. Holmes

without the consent of the author or copyright owner. Legal action will be pursued if this is breached.

<u>Disclaimer Notice:</u>

Please note the information contained within this document is for educational and entertainment purposes only. Every attempt has been made to provide accurate, up to date and reliable complete information. No warranties of any kind are expressed or implied. Readers acknowledge that the author is not engaging in the rendering of legal, financial, medical or professional advice.

By reading this document, the reader agrees that under no circumstances are we responsible for any losses, direct or indirect, which are incurred as a result of the use of information contained within this document, including, but not limited to, errors, omissions, or inaccuracies.

Table of Contents

Introduction

Social networking sites have seen an amazing growth rate over the last couple of years. As long as the interface is easy and visually appealing, they are a sure-shot win for the founders. Not only is this applicable for websites, but the incredibly large number of mobile applications that have been recently developed. These make social networking more easily accessible and are constantly being improved to make them better for the user.

One such networking service that has a large number of users from all over the world is Instagram. You are probably already on it or at least know someone else who has an Instagram profile. Ever since it was launched, the service is used by increasing number of people each day. Everyone from the girl next door to celebrities uses Instagram to make their social media presence more prominent.

Here, you will learn everything there is to know about Instagram. From the basics of creating a profile to utilizing different features of Instagram is covered here. Once you are done going through all the information we have put together for you, you can efficiently use this service as simply as you do a pen.

Chapter 1

How Did Instagram Begin?

Mike Krieger and Kevin Systrom are the main names behind the invention of Instagram. It evolved from an application for iPhones known as "Burbn." The Burbn app was a little too complicated in its initial stages, but the duo worked on it. They finally made it simpler and called it Instagram. At first, Burbn could be used to check in, make future plans, and earn points, among other things. They then decided to remove all the clutter and focus on being good at one specific thing. Their point of focus turned out to be sharing mobile photography. Considering it was similar to an instant telegram, they renamed it Instagram.

Kevin Systrom actually learned how to code all by himself without signing up for any professional courses. He then went ahead and built the Burbn prototype in HTML5 and gave it to his friends to try out. Soon after he attended a party he knew prospective investors were also attending. Since they agreed to invest in his idea, he quit his job and went all in. Kevin then started working with Mike Krieger to build Burbn. They realized all the different functions they tried to include were making things worse. Focusing on sharing photos, they built a prototype. However, this did not work out well either. They made Burbn an iPhone app but it still wasn" t what they were looking for. De-cluttering everything, they decided to focus solely on the photos. Renaming this simple, yet ingenious photo-sharing version as Instagram, they launched it. And, as we all know, it became an overnight success.

Although the app was initially just for iPhone users, they released a version for Android in 2012. This got them many more users in just a couple of hours of the launch itself. Soon after, Facebook started negotiation talks to acquire Instagram. Within a couple of months, and with all the legal technicalities out of the way, the deal was sealed. The amount in question for this deal was estimated at about a billion dollars.

Later in 2013, Instagram also launched a version for Windows Phone. Instagram also added the "Direct" feature in December of 2013. This allowed users to send private messages, which was aimed at competing with other social messaging apps.

Over time, Instagram keeps updating and making changes to the app to make it better than it already is.

Chapter 2

Getting Started with Instagram

Although Instagram can be accessed online, it is a mobile platform. To get started, you need to download the application on your device. You can only do so on Apple or Android devices.

Once you download the application, you can register your own personal account. Once you register, you can access this account by entering your username and password on any device that has the application.

After you have registered, you can see to other details. For instance, you can set a photo of your choice as your profile picture and change this at any time. You can then enter all the other information required on your profile.

You will also see that there is an option of entering a bio on your profile. The limit is set to 150 characters. You can fill in this field accordingly or even leave it blank. You can also enter a website link into your bio. Other users can click on this and go to the webpage by clicking on the link directly.

The account will be on public display for all Instagram users by default. However, you can go ahead and change it to private as well. This will require you to approve the "Follow" request any user sends you if they wish to view your profile. This privacy choice depends entirely on you.

Notifications

To set how notifications appear on your device, go to the notification center in your settings to configure. This will allow you to control if notifications appear externally or only when you are using the app. For external notifications, you can select Push Notifications in the "Settings" tab of the app. You get notifications under the News tab while using Instagram. They appear as a heart or speech bubble, depending on likes or comments. The feed under News will show you the specifications of any notification while in the app.

The notification feature will help you see when anyone likes your posts or comments on them. You can also be   notified if they mention your username in a comment on any other post. If someone tags you in a photo, you will be notified as well. You will also be alerted if anybody starts following your profile. As we mentioned before, this follow request will first have to be approved by you if you set your profile to private. Notifications will also let you know if any of your posts make it to the popular page.

Social connectivity

Instagram can be used singularly and is definitely more than enough. However, the whole point of social networking is to connect as far and wide as possible. You also have the option of connecting your Instagram account to other social media like Facebook, Twitter, Foursquare, Flickr, Mixi, Weibo, Tumblr, and VK. You can connect this by linking the profiles under settings.

Go to your Profile and edit "Sharing Settings" under that tab. You can then choose to which network you want to link your account.

This will allow you to share any Instagram post from your profile to the linked social network as well. You have the option to share or not, but the link is enabled if you choose to do so. If you do share it on any other platform, any other people you are connected to you at each platform can view the post. However, the privacy settings will still be valid on your actual Instagram profile.

Adding Photos and Videos

You can choose to share photos or videos on your Instagram account. This is the main point of setting up an account on this service. What you upload will make a huge difference in many things, the details of which we will discuss further later on.

First, let's take a look at exactly how you can upload photos and videos.

The first choice is to take a picture through the camera button at the center of the panel. This connects to the camera of your device. You can now take a picture or video through this connection.

The app also allows you to shoot or upload videos. These have to have a minimum length of 3 seconds and can extend up to 15 seconds. When you shoot a video, you can click on the record button to start. If you want a continuous segment, just keep your finger on the button the entire time you are shooting. If you want a video containing multiple different clips, just remove your finger at the places you want to stop.

The second option is to upload something that is already saved on your camera roll. Although you previously had to crop your images to the square that Instagram required, things have changed. You can now use pictures with any ratio without having to cut or add to them. Once you choose what you want to upload, you can go ahead and edit if you choose. Then all that left to do is to upload.

Editing

Once you select your photo you can choose to use it in its original form or, better yet, enhance it with some editing. Instagram has some amazing filters that can change the look of your image in different ways.

You can choose to use the readymade filters that are built on the app or do it yourself. Utilizing features like the brightness or contrast make the picture look much more professional. Tilt shift is another great editing option that will allow you to give your photos an altered depth of field. You can choose between the horizontal bar or circle when you use this and rotate your picture for the best effect.

The lux effect is like an auto-enhancing option. It enriches the image when you click on the sun icon. You can increase or decrease the depth of this as well.

You can also edit the pictures on other photo apps and upload onto your Instagram profile. The choices are nearly unlimited when it comes to this.

Filters on Instagram

The service has ready-to-use filters for any pictures you want to upload. You can choose to use the original photo that you click through the application. You can also upload pictures you have previously stored on your device. As opposed to using the photos in their original form, these in-built filters can help you enhance or modify the look of it. The following descriptions of the Instagram filters will help you get a better understanding of how each filter will modify your original image.

Normal: This is the primary choice of keeping the photo as it is.

Lo-fi: This filter helps to enhance the warmth of the temperature and increases the level of saturation. This will, in turn, make the shadows more prominent and enrich the colors.

Earlybird: This popular filter gives the picture the look of older days. The tint is sepia and the warmth of temperature level increases.

Amaro: This puts the focus on the center of the image and increases light in it.

Rise: The light on the subject is softened and a glow effect is placed.

Inkwell: The image is converted to black and white with no other modifications.

Sutro: The edges of the picture are given a burnt affect while highlights are increased. Any purple or brown points on the image are focused upon with stronger shadows.

Mayfair: This adds a subtle pinkish tone and the center appears brighter due to the vignette effect at the edges. A border is added with a slender, black line.

Sierra: This gives the image a softer look by making it appear faded.

Valencia: This gives the image an antique effect, as the warmth of the colors is increased along with exposure levels.

Nashville: This is meant to give a picture a certain nostalgic feel by giving it a subtle pink tint. The contrast is decreased while exposure and temperature levels are increased.

Toaster: This makes the picture look aged by adding a prominent vignette and burn to the center.

Perpetua: This is a good filter for portraits, as it adds a pastel effect to the image.

X-Pro: This adds a tint of gold and increases the vibrancy of colors. The contrast is increased and the edges gain a vignette effect.

Cream: This adds a hint of both warmth and chill with a creamy effect.

Walden: This is yet another popular filter that adds a yellowish tint to the image and increases the level of exposure.

Aden: This increases the tint of blues and greens in the picture.

Slumber: This decreases the saturation level and emphasizes the black or blue hues in the picture. A haze is added to give it a dreamy retro effect.

Ludwig: This decreases the saturation slightly and increases the light.

These are some of the basic Instagram filters that are used. There are a few more, and you should also note there are often new filters added with app updates.

Sharing

The screen where all sharing options are displayed appears once you have finished editing the image. You can add a caption to go with the picture or video. Adding hashtags will allow the picture to be visible when people search for those particular categories. You can also mention specific people in the caption with the @ sign and then their username.

You can also tag people on the image. Another option is to mention the location of where the image was taken. You can then choose to share the picture on other sites like Facebook or Twitter when you upload to Instagram. You can do this before or even after uploading the image to your profile.

Instagram Direct

The Instagram direct feature is used for sharing pictures or videos privately. You can choose exactly who gets to view these. In 2015, this feature saw the addition of direct messaging as well.

The above-mentioned steps are the most basic way to get started on using this great social service that the world loves. As you read on further, we will explain everything in greater detail, as well as how you can get better at everything you do on Instagram.

Chapter 3

How to Make a Great Instagram Account

In order to get popular on Instagram, you should have a great account with amazing pictures, a few other tricks up your sleeve, and you're set. There are millions of users who have an account on Instagram and you need to put in some effort in order to be amongst the best of them. If you put up the right posts at the right time, you're quite likely to get popular in a short time. As you read on, we will guide you to create an Instagram profile that will be the envy of others.

> ➢ Your first step in creating an account is choosing the perfect username. Your name should be catchy and should catch the eye of other users. Try not to make it too long or have too many symbols in it.

➢ The second step is choosing the perfect profile picture. Putting up your own photo is the best option so that people will connect your profile to your face. Find a fun, good quality picture of yourself to use for this purpose or take a new one.

➢ You should now give careful thought to choosing a theme for your Instagram profile. It has been noticed that pages with a particular theme get more followers and likes on their posts. This is because the followers will know exactly what the page is about and follow based on their interest. You can also choose to post whatever you want outside of any specific theme that limits you. However, a themed page has more chances of becoming popular. For instance, if you are someone who loves food, make your account food related. If you are interested in fashion, focus on that. This allows users to have an idea of what they'll be seeing in their feed by following your account. In fact, your theme can have variations other than a subject area. Your theme could be a color, where all your pictures are black and white or have a pink hue. It could also be random pictures of different places with your dog in all of them. Get creative with your theme and make it uniquely yours.

➢ After you have decided a theme, start clicking some great pictures. Learn how to use your camera in the best way possible to get the best shot. You can go ahead and click a hundred different photos of the same subject to get the best angle. This is worth it even if you just end up using one image you think is perfect. Get a firm grasp on how the editing options in Instagram work as well. This will help you enhance your images and give them the best visual possible.

➢ Build a portfolio before your start working on getting more followers. There are a lot of tricks you can use to generate traffic to your profile but very few people will follow you if your profile is empty. So start by uploading a few great pictures that will make your account look good and make users anticipate better posts in future. Don" t upload a meaningless, bad quality post. Quantity is not what you are aiming for, but quality definitely is. Work on getting these pictures up and then get those followers.

➢ Find out what the best photo editing applications are. Download the ones that suit your purpose and theme. Utilize all the amazing options available. Different applications offer different tools to make your photos better or more creative. Some will let you adjust the brightness, saturation, etc. Others will allow you to insert text in different typography on the photo. Use these for posting great pictures on your Instagram account.

➤ Give some thought to the captions. You can choose to have short or long captions. However, these should be to the point and not some boring narrative your followers lose interest in reading halfway through. Your captions could be inspiring, funny, sarcastic, or just about anything that inspires some type of emotion. They should, above all, relate to the photo and not be completely out of context. You can tell the story of your picture using its caption. This will help your followers relate more to your content.

➤ Be personal in your posts. After all, your profile is entirely yours and should be distinctive from others. Other users will choose to follow you only if they feel your feed is distinct from other run-of-the-mill accounts they have no interest in. Your followers should be able to feel a connection with you. It doesn" t mean that you have to have some kind of emotional outburst in every post. Getting personal means the user feels that they are seeing bits and pieces of you in your photos and not just a random cup of coffee. Your coffee picture should narrate a story of whom you are having it with or where you like it best. That personal touch can make a big difference.

➢ Link your Instagram profile to any and every other social media accounts you have. This could be your Facebook account or any other blog you have as well. This will help you generate a greater number of followers from there to your Instagram. If you have a huge following on Twitter or any such network, just update a post and let them know you are now on Instagram as well. Most of them will definitely follow through, if not all.

➢ Make your Instagram site unique as compared to any other social media account you have. Give your friends and other users a reason to follow your Instagram account specifically. Why would they do this if all your content were the same on every online profile that you have? If you have a popular Tumblr account, your followers would follow you on Instagram as well if they felt there was something more or different to see on Instagram and not just the same thing in a different format.

➢ Follow other users and like their posts as well. Other Instagram users usually want the same thing as you, even if they aren" t going about it the same smart way as you are. However, liking the posts others put up will adds a lot of brownie points for you. Find other people" s accounts where the posts are similar to yours. See who follows these accounts and then follow them. Once you appear on their news feed, they will probably check out your account and follow you back. You should also like the posts of the people you follow or else you will just look as if you are a ghost account. You can also choose not to follow these people and just like or comment on their pictures. This will also get their attention if you don't want to follow as many people as you want followers. Your comments will not only be noticed by the owner of the profile, but also other followers of the same person.

➢ Interact with your followers and other users. Comment on pictures other users put up. Most people appreciate comments more than likes on their pictures. You can be among the select few who make this effort. This will definitely make the user interested in your account and maybe make them follow you as well. You should also make it a point to reply to all the comments people leave on your posts. This will encourage them to continue commenting and liking your other pictures. It also builds a loyalty following for you because of the interactive effort you made. This will help build your own Instagram community.

> Use hashtags to direct people to your photos even if they aren't following you. The main way in which Instagram users discover each other are by searching for tags on categories that interest them. You should also pay attention to which hashtags are trending during that period and use them on your posts. This will allow more likes to be generated on your post while that particular tag is trending.

> Contests are yet another great way in which you can engage with other Instagram users. You can also hold your own contest at some point on your page when you have a lot of followers and a particular theme to your page. This will, in turn, direct more traffic to your Instagram profile and allow you to gain more followers.

➢ Do not post too many pictures too often. Your followers don" t want just your posts to flood their feed. Make sure to pay attention to quality over quantity. This will see a much more positive reaction from the user base. Post one good picture a day instead of ten bad pictures that have no point.

➢ Make an effort to find out what kind of posts people like more often. While you should stick to your unique theme, you need to give this some thought if you want a huge number of followers in a short time. Posting such content will only work to your benefit. For instance, most users have noticed that pictures taken at the beach or have sunsets are very popular and get a large number of likes as compared to others. By uploading such content, you will surely see more followers on your account faster than if you posted other pictures all the time. You should keep posting what interests you, but also make a conscious effort to put up pictures that interest the audience you want to keep engaged in your profile.

➢ Yet another thing that you should keep in mind, your profile should be public. The privacy setting should not be switched on to the private account tab. This will definitely not work in your favor if you want more number of followers on your account. It will actually make it harder to attract new follower when they can't even see what your profile contains.

➤ Give a moment to write a good bio as well. You can make it short, yet effective in order to gain people's interest. Make the content meaningful or catchy. It can be a brisk description of who you are or what your interests are.

Chapter 4

How to Get More Likes for Your Instagram Posts

The most obvious part of using Instagram and posting on it is that you want to share. And the positive feedback you look for about your posts is gained through likes. Everyone on Instagram wants as many likes as they can possibly get for their posts. This applies to someone running a personal account, bloggers, or even businesses. The more the followers and likes on your account, the better it is. Nobody likes if their post hardly gets five likes after having uploaded it ten or so hours ago. It would definitely be a little, if not very, disappointing. So, let's take a look at exactly how you can actively make an effort to get more likes on all your Instagram posts.

- The most obvious step is to always link your Instagram with your Facebook account. This will help all the people on your Facebook list to see you have an account on Instagram they can also follow. You can either share a picture from Instagram to Facebook or even actively ask your Facebook friends to follow you on Instagram. These are usually your first followers when you make an account. Get as many followers as you can by linking to all your social media accounts. This automatically gives you an increased number of likes when you post something.

- One of the first things you need to understand is if your post are of good quality, people will obviously like them more. So, make a conscious effort to post some good pictures that will catch the attention of other users. Your followers will definitely click the little heart and other users might just start following you out of interest.

- Your posts should be personal and attractive to your followers. It should be something that would interest them and capture their attention. Your posts should also have a unique quality that makes users want to follow you in order for them to look forward to future posts.

- Using hashtags is the easiest way to make your post more visible and discoverable by other Instagram users. These tags should be relevant to the content of your post. People tend to search for tags based on their subject of interest. If they like your post, they will like it and also start following you. The next part of using hash tags is to use popular ones. There are certain tags that are more popular at certain times. Posts with these tags generate a lot of traffic and get more likes. Some examples of such tags are those that have the words "iphonesia," "nofilter," "instadaily," "webstagram," "photooftheday," etc. Using these tags on your posts will make it more likely to get attention from other users.

- The filters that you use on your image also have a certain effect on users. Studies have shown posts with certain filters get more likes than posts with other filters. In fact, pictures with no filter or the original image get the most amounts of attention and likes. Making the photo as natural yet high quality will make it the most popular. In case of those with the use of filters, Valencia, Earlybird, Hefe are amongst the most popular ones. Using these will make it more likely for your post to get a like than any other filters.

- Commenting on the posts of other users will make it much more likely that they like, comment, and follow your account as well. It just makes you much more interactive than all the other users and generates the interest of the user you're interacting with in your account as well.

- Get likes by spamming other accounts. Go to the explore tab and search for certain hashtags like the following: #likeforlike, #l4l, #spamforspam, #spamback, #likebackalways. Once you search for these, go to the profile of any user who has this hashtag on their post. Like a few or many of their pictures at once. This will make you appear on their notifications. You can also leave a comment on any one of their pictures, and then ask them to spam back or like back your pictures. This is one of the simplest ways to get likes on your pictures. Others who want the same are quite willing to return the favor. One of the easiest ways to generate traffic to your account is by liking hundreds of pictures in one day. At least some of them will follow you and like your pictures.

- Timing is yet another crucial factor in determining how many likes you can generate on your post. There are certain times of the day or days of the week when more people are active. In general, Mondays and Wednesdays have shown to generate a lot of likes on posts. The other thing to notice is when your specific followers are more active. Analyze this using certain other applications and determine when you should post. If you post at a time they don" t even see your post, you are hardly going to get any likes.

- You should also not spam the feed of your followers by posting too much at the same time. This will actually annoy them and might even make them unfollow. Less drastically, they will just choose to scroll down without even looking at the post properly. Instead, if you have a lot of good photos of a particular event or subject, just make a collage. A collage is an easy, clutter-free way of sharing these images with your followers. There are so many apps that allow you to do this in an easy and attractive way. This will ensure that you get likes without being a spammer.

Tips for getting more likes on your posts

➤ Put up selfies or pictures with more faces. Studies show that a picture with your face on it is more likely to get higher number of likes than those without.

➤ Don" t post too much, too often. Appearing on someone" s feed too much can actually make him or her unfollow you. Try not to post more than a maximum of one or two times a day. Posting too much will decrease the number of likes you get.

➤ The hashtag #nofilter shows higher chances of increasing likes on a picture, even when people are actually lying about it.

➤ Use photo-editing apps to get a wider range of effects for your pictures. You don't have to limit yourself to Instagram filters. A lot of great photo editing apps can make your pictures look like they were shot professionally. This definitely increases the number of likes on your pictures.

➤ Don" t add too many hashtags on your posts. This makes it look as though you are posting for the sole purpose of getting likes. Seven is a good number for increasing the number of views via hashtags. You can find out about the popular hashtags that would increase your likes on sites like tagsforlikes.com.

➢ Get vibrant shots in the right light. Sunsets or a sunrise shot often see more number of likes due to the play in colors.

➢ Taking a photo at the right angle can make a lot of difference. For instance, when a person is standing in front of you, take a shot at an upward angle from below. This makes them look proportionally better and taller.

➢ Hues of blue on your photos are shown to be more attractive. Those with predominant red hues show fewer likes in comparison.

➢ Develop your profile according to who you are and what you like. Forget about others and what pictures get more likes or any such things. Take time to figure out what subjects you like to take pictures of. Experiment with different types of editing to see what suits you. Your profile should be unique and different from others.

➢ Taking time to shoot the same subject a number of times will help you get a perfect shot. No   one gets it perfect at one go. Look for different angles or a unique type of theme for your pictures.

➤ Take time to look at other people's feeds. Give some thought to the feeds of people who capture your attention. See what is in their pictures that attract you and how you can try to use the same techniques for yours. Following these kinds of people will help you develop your feed quicker as well.

➤ Use the Iconosquare app to analyze your activity. It will help you to see when you get the most number of likes and what time most of your followers are active.

➤ If you already have a great shot without applying filters, then improve it with the brightness and saturation tools as well as the other available tools. These will enhance the image without changing it drastically.

➤ You don't necessarily have to use the front-facing camera for selfies. Using the rear camera will give you a picture with much better resolution without the dreary, grainy effect.

- Don" t post too many pictures of the same thing and flood your follower's feed. Just put them together in a collage and share at once.

- Participate in hashtag projects held by the official Instagram page or other popular pages to get your posts noticed. If your post makes it to these or the popular page, your followers are much more likely to increase overnight.

- Use hashtags to increase the engagement on your posts. The more the hashtags, the more number of people are likely to view it. Instagram allows you to add up to 30 hashtags to each post and you can use them all to your benefit.

- Don" t edit your pictures too much. It decreases the likelihood of getting more likes. The more normal your picture looks, the more people like it.

- Brighter pictures get more likes than those that are darker. The same image with higher brightness got more number of likes than when it was darker.

- Cooler colors engage more people than a warmer palate with red, orange or pink hues.

➤ Get a theme. Decide on a theme for your profile. Most of the popular profiles on Instagram show a particular theme with respect to each. Users will get an idea of what they can expect when they follow you and are more likely to do so if they like what they see. Consistency is the key to maintaining a theme on your account. The theme could be location-based, food-based, or even color-based. The point is, when the user sees the overall account, they should notice the theme.

➤ Captions are another focus point. The picture could be great and the caption will make it even better. Your caption allows your users to get an insight into the story of the picture. Your followers should look forward to the next story you will tell.

➤ Consistency and rhythm when updating is also important. You should not post ten pictures in one day and disappear the next week. Or upload every day for a month and disappear for the next six. This might make users stop following your feed or even liking it.

➤ Get involved with your followers. Check out their accounts and maybe like some posts, even if you don" t follow them back. Reply to the comments they leave on your posts. Show you appreciate their likes and build a positive relationship. This will gain you a loyal following on your account.

Chapter 5

Instagram for Business

Instagram evolved from being a platform solely for social networking to a possible source of income. Using the concept of sharing on this popular forum is using it in so many different ways. A lot of businesses flocked here to utilize their services and reach out to the user base of millions. From big brand names to small closet shops, you can find the whole lot amongst the thousands of user profiles who use Instagram for business. As you read on, we will tell you can do the same and get positive results.

Instagram itself is not a service that sells any product or services. The platform, however, can be used for the benefit of such businesses. Brands have been using Instagram to market their product and make their presence valid on this popular service. Smaller-scale retail shops can be found on Instagram where followers are given instructions on how they can purchase items. The possibilities for using Instagram for business are vast.

> To get started, first create an Instagram account for your particular business. It works the same way as any normal user profile but keep in mind this is a business-centric account as opposed to a personal account. Due to the huge boom in users making use of Instagram for business, the service started a blog called "Instagram for Business." This helps users in more ways than one to see how they can make use of Instagram for work.

> Once your page is up, start posting. Make sure your posts are done with the customer in mind. Think of what would interest your customer, while promoting your business or products. Keep a balance between fun posts purely for the pleasure of the follower and those that are focused on benefiting your business.

> Make your posts targeted at the customers you want. Your posts should have that certain "thing" that targets these particular users. If your business is related to some healthy food product, use this as

your theme. Post about health in any way possible and inspire these users with similar interests to follow you. Your account can't solely be about being a catalogue for your product. Develop a certain message and image for your business or brand. Align this according to the customer base you are targeting but don't look excessively commercial or business focused.

➤ Make your posts customer-centric. Users will follow you when they feel your posts will benefit or interest them in some way. If you upload random photos that are irrelevant to the customers, they won" t get engaged with your Instagram. Put up products in a way so your customer feels interested enough to give time to actually look at your post and then maybe go ahead and buy it.

➤ Your account should be a visual experience for any user who views it. They will then get an idea of what to expect when they follow your account. It should be aesthetically pleasing and interesting. Find creative ways to showcase your services or products.

➤ As you start posting, start working on getting as many followers and likes on your account. This is the main point of using Instagram for your business; you want to increase the user interface. There are a lot of different ways to do this and we will explain this in detail further on in the chapter. The basics are to connect to other social networks, use popular hashtags, and keep interacting with followers.

➢ Use the Video feature on Instagram for uploading short advertisements for your business or anything else that will engage your followers. These are short and effective in generating interest. Share videos via links on other social media as well to generate more views.

➢ Make a schedule for posting. You cannot post infrequently or too often. Maintain a balance and keep your Instagram active but not overactive. Your followers obviously want to see your posts but not ten times a day. This is where a schedule helps you. Keep posts ready to upload and do so at appropriate times. Keep note of when your users are more active and post during these times. Make sure the posts are of interest to them as well.

➢ Use other applications that will help your Instagram feed become much better. There is a huge range of different applications that allow you to enhance your photos, print them, analyze your profile, etc. Spend some time in seeing which apps will help you get the most out of your Instagram experience. Go through the accounts of other popular users and see what they are using as well. Use the analyzing tools provided by some applications to see which posts garner the maximum interest and gain more followers for your account. These posts can be further used for more advertising, since they become popular images that users will identify with your brand.

➢ Make your posts as attractive as possible. The images should be high quality and vibrant. Make use of the filters on Instagram. You can also utilize the hundreds of other photo editing apps available for this. Make the photo look as aesthetically appealing to the follower as possible.

➢ Use the different hashtags, contests, etc. that are trending to market your brand on a frequent basis. This will keep you on the grid with all the latest things that grab the interests of users over time.

➢ Interacting with other Instagram users and your followers is very important. You should not just expect them to like and comment without responding as well. Reply to comments and like their pictures as well. As you build a relationship with them, they will be loyal to show their appreciation as well. You can also work with other established users, like bloggers, to promote your page and return the favor.

➢ You should also show followers how the inside of your business works. You may have noticed how many brands have been putting up pictures of events where they celebrate with their employees. This shows a positive image of the brand or business and creates a good name for it. So, go ahead and post a picture of a happy employee at work on their craft. Your customers will be quite happy to buy that product. Put up the picture of a work lunch or dinner. Maybe upload a video of some product being made in a creative process.

Benefits of Instagram for Business

Instagram is definitely one of the most versatile social media applications. Aside from using it to share photos and videos like a personal blog, it has proved to be an immensely useful tool for businesses. It allows any business of any size to reach out to the millions of users on the app. Take a look at a few ways in which having an Instagram page for your business can benefit you:

✓ It is an extremely convenient way to show your products to a huge potential customer base. People need to be able to see what's offered in order to know if they want to buy it. You can share good quality images of your products or services on your account.

✓ Yet another way to personally connect with the customer base is to show them how your business works or how the products are made. Most people these days want more details about how the products they buy are manufactured or produced. Sharing posts related to this makes them have more faith in your company.

✓ It is like a free advertising tool for your products and services. You can come up with creative concepts on how to showcase what you have on offer to the users on Instagram. Show them how the products can be of use to them and why they should come to you before anyone else. Instead of paying for other forms of advertising, use your resources to make some amazing concept posts. These will automatically reach out to more people if they're of high quality content which will then generate more customers for your business.

✓ The fact that the customers will get something extra by following you on Instagram can be a great incentive. Give them sneak peeks into new products and launches.

✓ You can get more connected by showing customers where and how your business works. Share pictures of your employees and what they do at the office each day. Show people how exciting or happy your work environment is and creates a positive image that definitely will work in your favor.

✓ Another benefit is that you can connect with other influential users on Instagram and get them to promote your products as well. A few tips and tricks like these help you reach the maximum number of people that you possibly can on Instagram. Since these influencers already have a loyal following, their followers will also be interested in you. Celebrities on

Instagram can also help you out with the right incentives. So go ahead and make use of these.

Instagram analytics

The analytics involve tracking the performance of the account. This means you can see how well or poorly your posts are doing once you upload them. Getting a proper analysis of your performance will ultimately help you perform better in future. This is especially essential for any business accounts. You need to have a clear idea of what your customers actually take notice of and what kind of posts in which they feel engaged. This helps you make more posts along those lines, thus generating more traffic. The data analytic tools provide will help you see if you are meeting the initial goals you set out to meet.

If you are not, you can then make changes accordingly to move in that direction. Depending on your follower base, you can generate a weekly or daily report. The analytics help you see your follower growth graph. It will also help you take note of who your top followers are and how much people engage in your posts. One of the best parts is that it helps to identify which timings and days are the best for you to post. These are peak timings when most of your followers tend to be using the application and have more chances of actually viewing your profile or posts.

Chapter 6

How to Make Profitable Posts

Since you are a business on Instagram, your main objective is to increase your profit in the long run. Let's take a look at how your Instagram posts can be used for this purpose:

> ➢ First, upload pictures of your products in a creative manner, but make sure the product is the highlight of a high-resolution photo. In case of a business account, put more time and effort in to getting the perfect picture you know will appeal to the consumers.

> ➢ Use easy links to allow your followers a fast and easy way to purchase any product that catches their fancy. If you upload a picture of a specific product, add a link in the caption to direct the user to the main

website where they can purchase it. This makes it convenient for them and much more likely for you to make a sale.

➤ Don" t be crude and overly objective in your posts. Be subtle in your message to followers when you coax them to buy your products. The post should be attractive and have its own story. It should not just be a catalogue image of a product with the price. Make the picture compelling and the product look so good the viewer wants to go ahead and buy it from the link you added.

➤ Use the Instagram direct feature for interacting with prospective customers. This allows the conversation to be private and you can send pictures, video, or messages. It doesn't just have to be about sales. You can also use this feature to reach out to followers who are particularly active on your page. This makes them feel appreciated and also gains you loyal customers.

➤ Make sure you offer discounts, contests, and other offers in some posts. This will make more people follow your Instagram account so they can keep track of these offers. This is especially effective if the offers are valid only for your followers.

> Like we mentioned above, contests are a great way to market your brand and get many more followers. Hosting a contest or giveaway can be fun and engaging for your followers. This will bring more traffic to your account and work in your favor for a small price. Let's take a look at an example of how a giveaway is held. The post should clearly state a giveaway is being conducted and state its terms and conditions. You can choose to state what the gift will be or make it a surprise. The instructions in the caption should explain how people can participate. For example, they should follow your page, tag your account in a post of theirs, and tag their friends as well. You can then choose a lucky winner from this or add more conditions to your giveaway. It's quite simple and allows more people to view your Instagram account.

How to reach out to more users

Via Facebook

Linking your Instagram account with your Facebook page will make a huge difference. Facebook has a larger user interface and will generate more likes for the same post. You can also set up advertisements or campaigns on Facebook that make a certain offer exclusive for those who follow your page on Instagram. This will definitely make more of your Facebook fans or friends flock to your Instagram page. You can constantly show you are an active entity on Instagram and make people eager enough that they wouldn't want to miss out on your Instagram.

Via the Official Website

If you already have an established website, use it to promote your Instagram page. This site gets most of your traffic from your active customers. Generate this traffic to your Instagram account by promoting it online. Generate banners on your homepage to link your customers to the Instagram account. Entice them with offers and campaigns.

Via Email

Emails are a huge part of most online marketing strategies. Your business probably has an email newsletter you frequently send. Add links to your Instagram in these emails. Make your Instagram presence visible via your newsletters. Show your readers how Instagram is a prominent part of your business and that they could benefit from following you on that site.

Via Contests

Instagram contests have become one of the best ways to generate traffic to your account. Nearly every single business on the service has utilized this particular method of marketing, as they witnessed positive results for others who tried it. Hold an Instagram-exclusive contest to engage your followers and other users by tempting them with a small gift. Everyone loves a free gift, whether big or small. Sometimes they don't even care that it's something they wouldn't normally buy. As long as it's free, people tend to want it. One way to engage people is by telling them they need to like certain posts, share those posts on their own accounts, and tag the brand on it. This allows more people to view your Instagram presence. Since something like this is quite easy to carry out, most of your followers will go ahead and share. You can then choose a lucky winner according to the terms and conditions met by the contestant. These contests can be weekly, monthly, or completely random. However, you will surely see a larger number of followers generated on these occasions. Another smart way to make this work is giving your own products as a prize. If users like it, they will be paying for it the next time. So, you are actually generating more customers.

Via hashtags

Users search for images that they like via hashtags. Find out which hashtags are most searched for on Instagram and use them on your posts. This will make your posts appear on their search results and generate traffic to your page, if your post is attractive to the user. Popular hashtags are the most common way to get likes and followers.

Chapter 7

Marketing Tips for Instagram Business

In order to make your presence on the platform prominent enough to generate profits, you need to use the right tips and tricks. This chapter will tell you how you can use some simple, yet effective, marketing strategies for your business on Instagram.

- The first thing to always keep in mind about Instagram is that it is a photo sharing service. Hence, the main thing to focus on is your photos are high quality with great content. The better they are, the more likely they are to gain popularity with likes and followers. Use different apps and a good camera for getting such photos.

- Sponsored ads by Instagram are another great option for your business. Marketing yourself requires quite a bit of effort and you can only reach out to a limited amount of audience. Sponsored ads you pay for will make your advertisements visible to just about any target audience you want. Investing in these can be a good marketing strategy as well.

- Use your bio strategically and creatively. Put up a link to direct traffic to your main business website. Use tools to make this a short link that connects to your website. Also, try to make this landing page familiar for your Instagram followers. They should be able to identify with it and feel that the contents of this page are directed specifically at them. The link can also be changed from time to time to direct users to a specific page where a certain offer or contest is going on.

- Help out others and they'll return the favor. Like we mentioned before, promote other users on Instagram. These users could be anyone or any business. As long as they help to promote you and direct other users to your account, these relationships will work to your benefit. Return the favor and you will build a solid relationship that you both can benefit from. This is why you should reach out to other users who have a prominent presence on the Instagram portal. This is especially easy if the categories of your businesses are similar or complementary to each other.

- Use trending hashtags that are constantly on the Explore page. This will help your posts be more visible.

- Engage the followers or users who visit your page to build a better connection. After they follow you on Instagram, entice them to give you their email addresses or to follow you on other social media where you have already built a presence. This helps on all fronts. Emails are the easiest way to have a personal connection with your customers. You can constantly inform them of anything going on at your brand or send them weekly newsletters to keep them engaged in your business. Keeping in touch is a very important part in building this relationship.

- Pay for sponsored accounts to shout out for your business. This is one of the cheapest options for getting more followers and likes on Instagram. These accounts are solely built to help others who pay them to get more traffic in their Instagram. They have a base of thousands of followers whom they then direct to your account. It's a great way to instantly increase your prospective customer base.

- Get connected to real-time Instagram influencers. These people are very prominent on Instagram and have some of the largest number of followers. If you somehow connect with them and make them use your products, it's a sure win. These influencers usually use products and post about them in a personal way. This influences their followers to try those products out as well and will thus increase your customers or, at least, your followers. If the influencer seems interested in your brand, the followers tend to follow as well. Their post should have a positive review about your brand or business. By tagging your account in the post and caption, they link their followers to you.

Using Instagram advertising

As we have mentioned above, Instagram has an advertising option that you can utilize by paying for it. This will be a step-by-step guide on how you can avail this particular service for your business.

➢ You first need to connect your Instagram account to the Facebook account. The advertisement will be created with Facebook Power Editor. Under the settings tab in Facebook, you will see the Instagram ads option. Click on it and add an account. Here you will have to enter your username and password for your Instagram account. Confirm these details and click.

➢ Next, select the type of ad you want. You can select from three main objectives for your ads. They could be for clicks for your website, video views, or installing mobile apps.

- Choose an image ad that will have a normal photo with a button that will link the user for clicks.

- Choose a video ad that will also have a button but will be in video format.

- Choose a Carousel ad that has a series of images to look through.

Depending on the type of ad you choose, you can accomplish your goal.

➢ Set your target audience once you have selected the type of ad you want. You can select the locations, age, and gender of the people you want your ad to reach. There are more detailed options to limit your audience further but these are the basics that will help you reach the people you want. Using this targeting option properly can make quite a big difference.

➢ Create amazing content for your advertisement so it captures the eye and interest of your targeted audience. It should be creative, yet clear in its message. A great photo or video is crucial for a great campaign.

➢ Use your advertisements to their maximum benefit. Calls to action are a great way to engage your viewers in connecting with you after seeing the ad. You could tell them to tag people in the comments, tag your account in the posts, use a specific hashtag, etc.

➢ Attract your audience with a great ad and see how it boosts your business in more ways than one. An advertisement can be used to build on the brand image along with generating more sales.

Chapter 8

Things to Avoid on Instagram

All the above-mentioned content will help you build a strong presence on Instagram. These will help you whether you are building a personal profile or one for business purposes. However, this chapter will help you avoid mistakes many people make. The main point of this chapter is to tell you what you shouldn't do as opposed to all the to-do tips we have put together in previous chapters. Making some of these mistakes can be why your profile never kicks off and sometimes even gets shut down by Instagram. Your profile should have a positive image and not generate any controversy that affects you as an individual or makes your business suffer loss. Some of the things mentioned below are just silly mistakes you can avoid, while some might turn out to be serious offences if you don" t take heed.

- ➢ Don" t upload pictures of the same subject from ten different angles in a span of two minutes. This will make your followers annoyed and might make them unfollow you.

- ➢ Don" t upload low-quality, grainy pictures or videos.

- ➢ You should also avoid stealing content from the profiles of other Instagram users. If you are reposting, make sure to give the attribution due to its original owner.

- ➢ Don" t upload pictures of yourself involved in any illegal activities. This could be anything from taking or selling drugs to breaking and entering a place. The repercussion for these could be big if reported to the authorities.

- ➢ You should also try to avoid posting any pictures that are not your own. Don" t search for pictures from other online resources and upload them as your own.

- ➢ Don" t comment negatively on someone" s body. Body-shaming is seriously offensive and will get you a lot of criticism once people notice.

- Do not post pictures of random people without their permission. While you might feel they would be proud of having their picture posted, many people are actually reluctant of being in the public eye. This might even cause them to report you.

- Don" t ignore your followers. Completely ignoring their comments on your pictures will make them see you in a negative light. Try to interact as often as possible, if not at all times.

- Do not use offensive language when you interact with others. This will create a bad image for you and even your business, if you represent one.

- Don" t compromise your privacy and security when it comes to certain matters. Additionally, don't compromise anyone else" s. Do not give away personal phone numbers, addresses, social security numbers, credit card details, etc. Be cautious of who you give such information to, even if it is on direct messaging.

- If you are a business, brand, etc., try to remain neutral on political issues. If you post something that takes a particular side on any given issue, it could offend quite a few of your followers who see differently. Try to maintain a balance and a peaceful front.

➤ Do not do anything that contradicts the terms and conditions of having an Instagram account. Read through their policy carefully and make sure you avoid violating it.

➤ Don" t use services that make you pay for likes and follows. The apps you should be paying for are those that help improve your content by automatically generating a more popular profile.

➤ Lastly, don't use Instagram for any illegal activity or has the potential to harm anybody.

Chapter 9

Impact of Instagram

One in every seven people in the world is on Facebook. Whatsapp has 700 million users, which is almost a hundredth of the world" s population. Instagram is catching up in its number of users as well. In this world of shrinking distances, rapidly changing technology, and monopoly of social media, people have gotten more used to having a virtual presence rather than actually plucking up the courage and effort to talk to people in person.

As a result of all this, the display picture has become a symbol of ego and pride. It symbolizes what we stand for and represents our beliefs. We see people changing it at least twice a day, waiting in anticipation of the number of likes they might get, eagerly waiting to respond with a "thank you" to people" s compliments, and continuously refreshing the page to see how views its garnered. It shows how cool and happening our lives are, or how updated we are with current affairs, even though in reality we haven't seen the back of a newspaper.

For example, so many people are changing the color of their display picture in response to the terror attacks in France, but how is this truly going to help anyone? How will changing the display picture on social media make any impact, other than showing to the world you know what has happened and sympathize with it on a very superficial level? Similarly, a display picture with some semi-clad kids in the background shows how aware you are of poverty and child labor, but accomplishes nothing.

Nonetheless, changing one's display picture is the call of the hour and wherever need exists, markets have responded to it in order to fuel the inherent profit motive of man. We have so many apps on our mobiles that enable us to edit our photos. As if publishing photos that are nowhere near true representations of ourselves is not enough, we have all these beauty apps that help us to filter, edit, and actually transform our picture into something else altogether. We have always been taught to appreciate ourselves in our true form, not to lust after skin color, a slim body, or perfection. As long as we are comfortable in what we wear and do, it does not matter how our hair looks, how shiny our teeth are, or how pink our lips are. But it has become a subconscious trend for people to click pictures and immediately edit it on Instagram, or any similar app that enhances the original photo, stripping it of all its original qualities, and injecting pseudo-perfection into it. It does make the photo look very glamorous and professional, but where is the reality? Do we always have perfect make-up and hair? I don" t think so. There are times we just scrunch up our hair, and eat ice cream with groggy eyes while wearing our shabbiest pajamas. But the impression these pictures portray to the rest of the world is that we always look good, have awesome friends, and have a happening life. It is a sad situation.

Most people also use these artificially enhanced images on dating sites. Imagine the harsh incredulity when that person is not recognizable in real life because they have put such a different image online. Virtual reality is indeed dangerous because it allows us to be something we are truly not without accountability. Plus, everyone is doing it, so why not jump on the same boat? It creates a false illusion about oneself that is never healthy. It moves us farther away from reality, because we do not have perfectly kohl-lined eyes, a full red pout for lips, or high cheekbones. The sooner people realize this, the better we will all be.

Take India, for example. Dark skin color has always plagued people, leading to launch of beauty products like fairness creams and face wash. Most of the young girls use editing apps to make themselves look fairer, or to lighten their skin tone. What good is this doing to their self-image? The belief that fair skin is beautiful is being entrenched by such actions and social media propagates these actions. Some intervention is required to put an end to this pseudo-projection of self.

Everything the world has two sides, like a coin. Social media platforms like Instagram have been criticized due to the unrealistic goals certain people set in order to make their lives or appearances seem perfect. However, there are many others who have been using the same platforms to work towards making a positive impact in the lives of others. It is important to acknowledge and accept this aspect of it as well.

The first benefit of this amazing application definitely has to be its actual purpose: to easily share and connect with friends in any part of the world. It is easy, fast, and very user-friendly no matter who you are, where you are from, or even how old you are. The application allows you to take high quality images of just about anything and share it with the rest of the world. It is up to you to choose if your account is visible to the world or private. While a public profile with detailed personal information might have a negative result if an ill-intentioned person comes across it, the account owner should just be a little more cautious in what they display. Otherwise, it is a great way to share the little and large joys you experience in your everyday life.

Instagram is very effective in helping people to make a positive change in their lives. Instead of focusing on how beautiful and unrealistically perfect a person looks on their Instagram account, there are better aspects to notice. There are literally hundreds and thousands of profiles where health and fitness are the focus. These pages could be personal profiles of a particular person who is very enthusiastic about these topics or a profile where images of other fitness-focused people are shared.

The way to let this influence you positively is to try not to look like someone in that picture. You need to focus on how much healthier and happier the person will be if they put in a little effort towards his or her personal health. You need to make similar efforts to put in those hours of exercise and eat a healthy diet. This will help you get a great body that is uniquely yours and does not have to look like theirs. The goal is not to look like another person but to be the best you that you can be.

It is also a great platform for just about anyone to take pictures and share with everyone to show their creativity. It allows you to capture memories and also make some sort of art out of each picture that you take. Over time, more and more people on Instagram have been experimenting with all the different concepts that can be tried out on Instagram. This helps them create a unique online presence for themselves where they can freely express themselves.

Business on Instagram is yet another amazing aspect that has seen positive results. Quite a lot of profiles on Instagram are created for the purpose of business. Various brands, big companies, and small setups alike realized the potential of Instagram to reach out to more people. This is why they create an account to showcase their products and services on Instagram. It allows them to reach the millions of users on the app and also have a much more personal interaction with them.

As you can see, the same app can have advantages that outweigh the disadvantages, depending on the viewpoint of the person. The impact of this one simple application on society has actually been enormous.

Chapter 10

How to Edit Pictures for Instagram

The focal point of your Instagram account is the photos you upload. Whether a user sees one single post or an overview of the entire profile, their impression will be based on the quality of your content. Hence, you need to make an extra effort to first dish out photos everyone will love and envy.

For this very purpose, we will give you some tips and tricks to edit your pictures like a pro and make some attractive content for your account. Nobody on Instagram actually wants to look at blurry pictures taken in the dark where hardly anything can be discerned. So, let's see how you can avoid this situation.

> ➤ First, your backdrop is important. If you want your focus to be on a particular subject, keep a   clean

background and few other things in the picture. The focus should be on one subject. Using white walls or any solid-colored backdrop works well.

➢ Secondly, try to take as many pictures as you can during the day. Natural light is your best friend and will make your photo turn out well even before you start the editing process.

➢ Once you have taken your pictures, use the editing tools in Instagram or in other photo editing applications. There are so many to choose from that you will be spoilt for choice once you get started. Let's look at some of the basic corrections you can make on the image.

- The lux option in Instagram allows you to automatically make a balanced improvement in the picture without you individually fiddling with details like brightness and exposure. The slider can be used to adjust just how much of the effect you want to change in the picture.

- Always avoid taking pictures in dark areas or in dim light. Try to get as much natural light as possible or look for other sources of light. For pictures taken at night, download an application that will allow you to get better results when you take pictures in the dark.

- The tilt-shift option is yet another option for improving your picture. You can tilt the plane of focus and adjust which part you want to blur out a bit. On the Instagram option you can choose between a radial or linear point of focus. You can also physically increase or decrease the area of this focus using your fingers.

- Straightening images is another option offered to users on the Instagram app. You might have taken a picture at a slightly wrong angle. With this option you can actually straighten the image out. The grid in the app allows you to have a clear look at how straight your lines in the photo are and help you do a better job of making sure it becomes straight. The sun can now set on a perfectly straight horizon thanks to this particular hack.

- Making collages from your pictures is an easy way to avoid spamming your the feed of your followers with too many posts at once. You can download the Layout from Instagram app for this or use any other available photo collage app. There are so many layouts you can choose from in order to get that perfect picture grid you want for your Instagram post. Collages can be fun and creative, or clean and sophisticated. There are many options for you to play with until you are satisfied with the result.

- Adjusting or changing the tint of the overall picture is yet another option. You can make a normal picture a black and white shot. You can adjust the depths of the black, whites, and greys in this as well. You can also adjust the tint to make it more red, blue, or even yellow. You can then adjust the extent to how much the photo gets tinted in the manner you choose.

- Using all the inbuilt filters in the app is obviously the quickest and most popular choice for users. The filters available in

Instagram are actually quite great and can make your pictures look much better. You can adjust certain things in the picture even after you add a filter to it. For instance, you may add the Sierra filter but then reduce the warmth or saturation in the picture using other tools. Readymade filters from many different applications can help you to make you Instagram posts look quite stunning and appear as though a professional photographer took them.

Tips for some great Instagram posts:

- If your subject is food or your meal, the best trick in the handbook is to stand on top of your chair and take a picture from above. This gives a great photo of your meal in a clear and much more sophisticated way than taking close-ups of the food. We can" t assure you that you will look sophisticated while standing on a chair with your phone in hand, taking a photo in a restaurant. But it is definitely worth that momentary embarrassment compared to all the Instagram glory you will be showered in. Make sure you brighten up the picture, though, since most food joints have dim lighting. Increase the contrast, sharpen your photo, and you're good to go.

- Some unique walls are a great prop for your pictures as well. These could be plain white or even colored. The walls could be modern and clean or some part of an old brick building. Just make sure it is uniform and go pose to your heart's content, then watch the likes pour in after you post!

- Minimal themes are yet another popular way to get attention. The point is to have the subject take up a small bit of space in the picture with nothing around it. Make the picture is bright, yet balanced.

- Travel shots are a win-win on almost any given day. You could choose to take photos of a cityscape or of perfect scenery. For cityscapes, take the shot from the street and show a particular scene. Make sure the clarity level is high. In case of scenery, try to focus on one thing like a tree and make it the focal point of the picture. Adjust the brightness and contrast accordingly.

Different applications that help to improve your photos:

- VSCO Cam is an application that has a lot of loyal users. Many of them use it for every single post they upload on Instagram and it actually serves as their theme. You can customize the details in the picture and also choose from the great filters in this app.

- Overgram is another great app that allows you to add text in various beautiful fonts on top of the photo you want to upload. This is especially nice for pictures where you want to add quotes to give an insight into the subject.

- Squareready is an amazing lifesaver for those who like to upload all their pictures as a whole without cropping anything out. It allows you to fit the picture into a square that does not cut out any detail.

- Afterlight is also a good app that has many filters, all of which add vintage effects to your photos. There are a variety of filters to choose from and they all add a special something to your image.

- Snapseed is a very popular app and why would it not be? It gives you many options to edit the photo in a number of ways. There are so many tools on it and the end effect can make your photo look like a professional took it.

➤ Facetune is also a good app for when you want to upload pictures of people and they would rather not display some tiny imperfections. This app helps you make these minor changes like removing a pimple or even smoothening out a wrinkle.

➤ Hipstamatic is an app that lets you add a retro effect to your photos with its great filters.

➤ Everlapse is a video app where you can take photos and make them into a video to upload on Instagram. This is a great option to tell a story or just to upload a lot of pictures from the same day in a single post.

➤ Superimpose is yet another quirky app which lets you change the background of the image, swap faces, or even blend two images together. You can use this to make some fun photos or just some really cool posts.

Chapter 11

How to Get More Followers on Instagram

One of the basic goals of establishing a successful presence on Instagram is to get as many followers and likes for your account as you possibly can. It actually is not all that hard and depends entirely on how you use the application. The following tips will help you gain more likes and also increase the number of followers from the user base on Instagram:

> ➢ First, you need to start liking as many photos or videos other users post. This is one of the best ways to make people at least view your account and then like your pictures and even follow you. Like the posts of everyone who comes into your newsfeed, as this will make them feel inclined to return the favor as

well. Now, like or comment on as many popular pictures as you can. You should also like more posts with hashtags similar to the ones used on your own posts. This will generate the right kind of traffic for your page.

➢ Yet another easy way to gain followers is to tag your own posts with hashtags such as #followme, #followback, #followforfollow, etc. Other users who use the same tags and want more followers as well will see your posts and follow you. Now, search for posts with these hashtags, go like their posts, and follow them back as well.

➢ Comment on as many photos of other users as you can. This is much more personal than just liking their posts. People are also more likely to notice your comment on their post than your random. Commenting is much more interactive and effective, since you are adding a personal touch.

➢ Captions are a great way to engage your followers and other users who view your posts, thanks to the relevant hashtags you add. Make the caption count and try to engage the viewer. You can add a story behind the picture. For example, you can add   a question asking users about their views on something. This could be anything from asking them how your hair looks in a picture to how useful they found a particular product of yours if it is a business account. In fact, if you ask for reviews, it actually shows users you are open to receiving real feedback. Your caption is also crucial when you are holding a photo contest or something similar. These are all call-to-action captions that demand users become engaged with your post.

➤ One other way to keep followers and get new ones is to post consistently, yet not too often. Being overactive on Instagram will label you as someone who spams the feeds of users all the time. Nobody wants to see multiple posts from the same account in the span of three hours or even a day. Try to limit the number of times you upload something and ensure that these are actually quality content.

➤ Use the analytics tool and find out when more people are active on Instagram. Choose these timings to upload your pictures since this will ensure that you get more likes and even more indirect views. This will generate more followers on your account as well. There are many apps, like Iconosquare, that help you for analyzing purposes and will work to your benefit, especially if you take your Instagram presence seriously.

> Shout-outs are another easy way to get a lot of followers at once. This can work to benefit two people at once. Get in touch with a user who has a lot of followers and ask them to give you a shout-out on their account. All they basically have to do is share a post with you tagged in it and then ask their followers to follow you. Try to get them to post a picture that you shared on your account and to tag and mention you in their caption. If the caption is interesting and personable, it works out even better. It should not look as though you are desperate for more followers. Similarly, you can return the favor in kind. There are a lot of accounts that actually just do shout-outs for people who want a large number of followers for their account. Some of these promise quick results if you are willing to invest monetarily as well. There are certain apps and websites that also help you to get a large number of followers if you sign up with them.

> Now, if you let go of all those quick and easy methods to get a lot of fast followers, it all comes back to you

and your account. Making a great Instagram account in and of itself should be your main goal. There are quite a few tips and tricks that actually help get more followers for your account. One of the first things users get attracted to is a profile that has a theme. If you stick to a particular theme then users know what to expect if they follow you. Stick to certain subjects and those who are genuinely interested in those will more than likely follow you and like your posts all the time. Try to choose the theme based on something that genuinely interests you. This will drive you to further stick to that theme. It could be anything from fitness and food to clouds or the color pink. However, choosing a theme is not a necessity. Just make sure that your content is as good and high quality as possible.

➤ A profile picture and catchy username are yet another aspect that can work to your benefit. Choose the best profile picture of yourself or one with your brand name on it if it is a business account. The profile picture needs to be related to your accounts theme if you have one. The username should also be somehow related. The users should be able to connect with it and associate your username when they think of the subjects your account deals with.

➤ Use your bio section carefully and cleverly. It can give other Instagram users a quick view into who you are or what your account is about. Use some small phrases or a few appropriate words to describe what your account is exactly about. You can even add in the link to any other website that you own or manage in the bio section.

➤ Another great way to get more followers is to follow other people. Find people to follow across Instagram through different methods. First, try to follow all those who are on your contact list and have an Instagram account. You should also link your account to Facebook and follow your friends from there that have an Instagram account. The other users you choose to follow should have similar interests. Try not to follow too many people at once. You should also try to keep the number of people you follow below your number of followers. Users tend to shy away from following those who have more "following" than "followers."

➢ You can also get a lot more followers if you link your Instagram account to any other social media that you use. For instance, you can link it to your Facebook or twitter account. You can also share a link to your Instagram account on your blog or the company website. This makes more people see that you have an Instagram account and they just might follow you. You can also sync your posts with sites like Tumblr and Foursquare. Just tick the box beside the options for sharing. Every time you post something on Instagram, it will automatically get posted on those accounts that are linked as well. You can also choose to do this for specific posts and keep the sharing option turned off for other posts.

➢ Hashtags are yet another important part in your Instagram experience. Users search for tags with words related to their interests, which is how they discover new accounts they might choose to follow. There are two types of account users when it comes to how they use tags. Some like to use as many tags as Instagram will allow them to put per post. They then choose to use all of the most popularly searched words for their tags to generate traffic to their post. Since there are so many users who do this, it might actually work to get likes. These users use all kind of tags such as #love, #likeforlike, #likeforfollow, #followforfollow, etc. However, there are the other types of users who find such tactics desperate and inappropriate. These users would rather use a few choice words for their tags that are actually relevant to their particular post. This drives users who have similar interests towards their page and they are the ones who are more likely to follow your Instagram account if the content is what they prefer for their news feed. For such users, make the hashtags specific to what type of traffic you want to generate towards your page. As long as these users like what you are posting, they will look forward to more of your posts and follow your account.

➢ Like we mentioned above, tags play a very important role. If you would rather have a lot of followers fast and in a short period of time, use the popular tags. Popular tags are keywords that are searched for most commonly at a certain point in time. There are apps and analytics that tell you which words are being used for hashtags the most at any particular moment. You can then go ahead and set these hashtags to generate the maximum amount of traffic to your post. Using popular tags will make it possible for more people to see your post when they search for a particular tag. Even getting them to see you picture might just make them click on the post and like it. Here as well, having a post with good content will help. If your picture is attractive then it will catch the eye of a user, regardless for what they were searching. The tag does not have to be related to the subject of your picture. You can use just about any word more users of Instagram tend to search for. Although many people look down upon this practice, it actually does show results.

➢ Tagging the location of your pictures is also another option to generate traffic to your posts. If the location is a popular place, there will be people who will search for other pictures of that same place. This will help direct those users to your account and post. Geotagging is a great personal touch to add to your pictures and makes them more authentic.

➢ When it comes down to it, your content is what actually matters from the start to the end. The Instagram app is all about photos. You need to take pictures of things other people will find pleasing. The photos should also be as high quality as possible.

First, choose a particular subject for your post and take an attractive picture. If you have a theme, then stick to that theme. Take a number of pictures of the same subject from different angles to get the best shot. Then edit it to improve it further. Instagram has built-in editing options you can use to your benefit. There are also many other applications that allow you to edit photos and videos quite nicely. Use these to your advantage and improve your feed. Even if your subject is something common, the way you present it can be different. Make it look good in natural light and focus on one particular thing. Anything around the subject should serve as a background. Take pictures from different angles and with a unique perspective to get that one perfect shot for your account. This is what will make people attracted to your account and make them follow you.

Chapter 12

How to Hold an Instagram Contest

As we've mentioned quite a few times already, holding an Instagram contest can definitely work to your benefit. This contest concept has shown positive results for nearly any business on Instagram, as it generates lots of customer interest. It is a quick and effective way to engage customers on your account and get even more followers who get excited about the subsequent rewards you offer.

The best part is there are quite a few simple, yet amazing types of contests you can hold on your Instagram account. Read on to get a better idea of each.

Types of Instagram contests:

- One simple and random type of contest is where you ask your users to follow and like a particular post.

You can randomly choose a winner from amongst all the users who do so. Give a deadline until the contest is valid and announce the winner after that. The prize you offer can be your own product and if they like the prize, you have a returning customer for sure. This type of contest is a very simple way to gain more likes and followers at the same time.

- Another type of contest is where the users have to post a particular type of picture and use a specific hashtag relevant to your contest. The theme could be anything from asking users to post a picture of him or her or wearing something like a pink dress. The user needs to upload the picture and tag your Instagram account. They also need to use the specific hashtag that you create for your contest. The next part is that the winner with the best picture and maximum number of likes on their post will be the winner. You can then choose the winner after the contest deadline passes and announce the result on your page along with what you will be gifting the winner. It's a simple and effective concept where the pictures are easy-to-take selfies. Most customers find such contests appealing and will participate every time you hold such events, for the simple reason it's easy and they just might win.

- Yet another concept for those with an actual offline business is an in-store photo contest. This works for brands or shops that have an Instagram account and an actual setup customers can walk into. Your contest rules ask them to click pictures in your actual store and then upload the picture on Instagram. They have to tag your account and use the relevant hashtag. This allows you to attract more customers into your store via foot traffic, as well as engage them online. The contest timeline can be range from a week to a month. A weekly period gives a faster call to action and results.

- You can also allow contest participants to enter through your website as well as Instagram. Giving more options means you get more people who might or might not be using other platforms. Just make it clear they can use the option of Instagram or the company website.

- Yet another type of photo contest is where the customer is required to upload a picture with your products. The post needs to show how they use your products and also include a contest-specific hashtag. This way, other people get to see more real people actually using your product. A lot of potential customers can be generated this way. The users will also make a personal effort of making the images appealing to their own followers as well as to those judging the contest. It is also very brand-related and is a good marketing strategy. You can then offer up your products to these participants for their loyalty value. One example of this is be if you are a makeup company. Ask the users to create a particular look with your products and show what they have used clearly in their entry. It's amazing to see how many people get interested in such a contest and might use this as an excuse to go ahead to buy new things in order to participates.

As you can see, there are a lot of easy and creative ways to utilize Instagram contests to your benefit. Everybody loves free gifts, so entering a simple and fun Instagram contest is quite appealing to most people. While you part with some simple stuff you already have, you generate much more traffic that will pay to get the same stuff. Users don" t have to spend a penny to enter the contest and you earn more than tenfold of the one penny you part with. So go ahead and try holding a contest from your account. The results are more than likely to be positive.

Use the following tips to hold a successful contest on your Instagram account:

➤ First, set a proper goal or objective you want to meet with your contest. Everything you do needs to have a purpose and positive result.

➤ Now, think clearly about the target customers you want to reach through your contest. Your contest specifications will depend according to whom you want as your participant or future customer.

➤ Depending on what your goal and target market is, decide what you will be giving away as a prize. It should be something that will make your followers and other users want to take part in the contest. The prize should also be decided upon depending on a budget that you should not cross. Just make sure that the incentive is such that it engages your target crowd.

➤ The post you put up to announce your contest should also be appealing and attention grabbing. You can also show the customers the prize they will be getting if they win. Make it as attractive as possible.

➢ According to the target customers whom you want to reach, decide on what type of contest you are holding. You already know there are different types of contests you can hold and modify them creatively. For instance, if your brand sells tea, ask users to upload selfies of themselves drinking from your particular brand of tea. Similarly, decide on a particular unique theme for your contest.

➢ Manage the contest by asking all the participants to use a specific hashtag you create for your contest. They should also tag your account in the post so you can keep track of all the entries. Make sure you set specific rules for any entries so only users who follow these will be eligible to participate in the contest. One of the basics is that they actually follow your Instagram account.

➢ The hashtag for the contest is actually quite important. It should not be something people commonly use. That would make it difficult to determine who is actually entering your contest. The unique contest hash tag makes a community just for your participants so they can view each other's entries easily.

> Set a deadline for when the contest is running and until when entries will be accepted. Small weekly contests have been found to be quite engaging and productive. It keeps generating traffic and the investment does not have to be over the top either. If your prize is much bigger however, you can allow a longer deadline to make the users actually work for it. In case of such big prizes, you might choose to have terms such as the picture with the maximum likes wins the contest. This gives the participants more time to work towards the likes and allows more users to view your brand name as well. Make sure you also clearly mention on what basis you will be choosing the winners of the contest in order to avoid any confusion or conflict.

> Depending on the results you see from your contest, determine how often you want to hold them. If you see good and quick results with a lot more new customers, it makes it feasible to have frequent contests. Otherwise, take it a little slower with a monthly or seasonal contest.

➢ To get more participants for your contest, promote it in as many forums as possible. Announce the contest on your website or any other social media in which you have a presence. This could be anything from Facebook, Twitter, a blog you have, etc. You can also make other popular account holders on Instagram announce the contest on their personal account. You can also use ads on Facebook or Twitter and other such places that will allow more people to see what's being offered.

➢ Once your contests are up and running, monitor it closely. Use tools like Google analytics, Google alerts, Twitter analytics, or other such third party options. These will help you easily and effectively monitor your contest until it ends.

➢ All you have to do when you choose the winner is contact them and announce it on your page. You can take it a step further by uploading a great picture of the winner with the actual prize that you sent them. These makes more users interested and looking forward to the next contest that you will be holding.

Chapter 13

Users on Instagram to Follow

Following some popular Instagram accounts can actually help you make your account better. There are a lot of your favorite celebrities who have personal accounts. You can get an exclusive look into their lives and interests through these accounts. There are a lot of accounts of users who concentrate on particular things like makeup or food. There are accounts for just about anyone's interests and you would be surprised at how much time people spend looking at these each day.

Popular Instagram Accounts

- Rihanna has over 20 million followers on Instagram. Follow badgalriri to get a look into her everyday life and insider looks into any new music she might be coming out with.

- Serena Williams might just be your female sports idol. Now you can follow her on her account, serenawilliams. Get a look into exactly how this confident female goes through her days.

- Ellie Goulding has an account under the name of elliegoulding and shows a lot of backstage images for all her loyal followers. There's a lot you can see, starting with how she keeps herself fit to all the people she hangs out with.

- Jennifer Lopez is yet another ardent user of her personal Instagram account. Follow J-Lo to see all of what she shares from her personal life. Everything from her music to family life can be glimpsed on this account.

- Food Porn is a great account to follow if you have an uninhibited love for good food. There are so many amazing pictures of food on this page that you will literally feel hungry every time you see a post they put up. So follow food_porn to see all the delicious food from around the world that you might want to eat.

- John Mayer is yet another amazing celebrity musician whose account you have to follow. He is actually quite active on the app and shows you lots of bits and pieces of his life and career. One of the best things you discover on his account is his amazing sense of humor. And, of course, there's so much of his music.

- Karlie Kloss is another active star on Instagram who gives you a behind the scenes glimpses of her glittery life. This model has risen to fame quite fast and shows you just how much she loves all of it. Not only do you get to be inspired by this gorgeous female, you will probably envy the unfiltered look into her world.

- Cristiano Ronaldo is definitely the stuff most female dreams are made of and men aren" t far off either. Get inspired by just how fit and focused this football star is. And you will obviously get to see just how much fun he is having as well.

- Zooey Deschanel has an Instagram account you definitely should follow if you" re a fan. You get to see a lot of her unique personality through her posts, and there" s also a lot about her show, *New Girl*.

- Miley Cyrus is a child star who hasn" t fallen from fame unlike most her counterparts. You get to see a lot of the new image this particular celebrity has created for herself. While her Disney days may be behind her, she is raking up some good music and quite a lot of drama as well. Her account will give you a fun and quirky look into her life.

- DogsOfInstagram is a favorite for all the millions of dog lovers around the world. You can get a daily fix by looking into puppy dog eyes every single day. A lot of users send in pictures of their adorable dogs and you can do the same to share some dog love too.

- Lauren Conrad is yet another personality who has a great Instagram account. She shares a lot about everything from fashion and beauty to her personal and professional life. This beautiful damsel" s account is one you won't regret following.

Chapter 14

Some External Tools You Can Use to Gain More Visibility on Instagram

A recent study showed that Instagram gives popular business brands about 25% more interaction and engagement with their followers than most other social media websites. You can target very specific audiences and literally create visibility for yourself by making full use of the options Instagram offers. Before you start looking at marketing options on Instagram, there are a few extra tools you need to take look at and learn how to use. In this chapter, we will check out these external tools and the features they offer to help you gain more followers in order to augment your overall business strategy.

Instagress

As your account keeps growing, it becomes more and more difficult to handle and track. You can't afford not to respond to followers, but at the same time, there become too many for you to manage. Big companies hire social media managers, but if that is not an option, here is a program that does the work for you! Instagress puts your Instagram account on something of an autopilot, doing all the work for you. But beware – it has to be used effectively or it could get you into real hot water, even going so far as to get you banned!

So what is Instagress? In essence, it is like an automaton that posts, likes, comments, and shares on your behalf. In other words, it does everything for you. Now, the problem is that if this „inhumanness' is detected, you could get banned – your account could get flagged or even deleted! So what you need to do is use it correctly. If you set it up right, you will be fine! Start by connecting your Instagram account to your Instagress: log in with your Instagram account name and password. You don" t need to worry about security; it doesn" t store your password, so your account remains safe and under your control.

Once you log in and see your dashboard, you can set Instagress up to work for you! For three days, you receive a demo account and a free trial – the only thing you have to do is click on the start tab for your trial version to begin! The countdown runs from that moment to the next three days, at the end of which you can pay if you like the program.

So how do you use Instagress? Start by heading to the main settings option, where you can set up your preferences. The first thing you need to do in this initial phase is to set your speed to slow under the Activity Speed tab. This way, you can let Instagress work for you for a while and examine how effective it is. Later on, you can speed up your posts once you've figured out how to customize it to your needs. This slow speed also ensures Instagram won" t flag you for a sudden outburst of activity.

There are a variety of options for you to use in Instagress; the „tags' tab lets you explore the different hashtags that are trending and target your relevant audience, while the „location' tab lets you target people from specific geographical locations. You also have a „minimum like" and a „maximum likes' tab which lets Instagress know how many likes should be on a post before you engage with an account – you don't want to be the first to speak, and neither do you want your voice to get lost in the din!

Here are a few of the key features of Instagress –

- *Advanced Filters* – Recently added, this feature lets you get very specific with your target audience. The settings here vary depending on your personal requirements – you will have to figure out your target audience and then set it up accordingly. It is easy to navigate so it shouldn" t be too much of a problem! However, if you narrow down on your customers list too much, Instagress may not have a very big list to work with and it may stop its activity. So, be careful when you narrow your target list down and keep updating it constantly!

- *Commenting* – As we said, Instagress comments for you. This means you don" t actually even look at the posts you're commenting on, since the software does. This is the area where you could flagged – you need to make sure your comment is not generic, stands out, and is still cool and catchy! Pick the „don't comment same users' tab to make sure Instagress does not give the same comment to different posts by the same user, or they will realize an automaton is doing your job. Set up different comments you could post and add hashtags and emojis to make them stand out instead of appearing generic. Go back and keep updating to make sure that you sound real, cool, and engaging!

- *Following* – Instagress helps you easily increase the number of followers. Pick on the „followers of (usernames)‚ tab. This means Instagress automatically follows people following an account you have specified; as you know, the best way to increase followers is to follow them. They will notice you and follow back; this way, you can do it in bulk and get a number of others to follow you in return!

- *Tagging* – We have already discussed how important tagging is to the visibility of any Instagram account, and in turn, a business venture using Instagram to promote themselves. One thing to be careful of is using tags spammers use. Sadly, there are great hashtags you may have to forego since they are used by spammers. Key in to Instagress the best tags you can think of and keep updating the list to make sure you are always hip. If there are tags you want to avoid, put them under the „blacklist‚ section so Instagress will not post them on your behalf.

- *Location* – A recently added feature, this lets you reach a target audience in a particular location, making your marketing strategy specific. Obviously, you need to know who your potential customers are and where they" re from. When you know this, Instagress can customize your posts to cater to their needs in particular!

- *Usernames* – Set up your Instagress to follow or engage with specific user accounts so that you can build a community on Instagram, thereby promoting brand loyalty. Follow them, interact with them, and engage them in conversation so you know where you stand with them and where you need to go in the future.

- *Blacklisting* – As we saw in the tags section, you can put up tags you want Instagress to avoid under   the „blacklist' tab. But this isn" t limited only to tags; any content or user you don" t want to get involved with can be blacklisted. This way, you can avoid inappropriate conversations and unwanted attention.

- *Autostop* – Obviously you cannot keep following more and more accounts. Given that Instagress is an automated program, it will go on following people unless you set parameters to stop after a point. For example, set it up to follow 800 people and then unfollow around 300 – this makes your account look authentic and real. Also, keep in mind Instagram doesn" t allow you to unfollow easily; you need to space it out, which means your account growth rate could be stalled. So, be careful in setting up your follows and make sure you look authentic at all times.

Ultimately, keep in mind that although it's super useful, Instagress is a program that has to be used with extreme caution. It is, literally, a robot, and follows every instruction to the letter. Instagress does not think or create on its own, so you need to be very specific in your instructions to make sure that you don't get flagged and look authentic. But if you set it up right and use it properly, it can be one of the best marketing tools out there to get you the most visibility for your business!

Schedugram

Schedugram is one of the most effective and easy tools that will help you manage your Instagram account. It is a third party application that is no way affiliated with Instagram itself, but it has, over the years, proven to be an excellent marketing tool for those using Instagram in order to gain visibility for their business ventures. It is cheap, reliable, and offers you a number of benefits that other apps performing the same functions do not.

So what is Schedugram? As the name suggests, its primary objective is to help you schedule and time your posts just right so you can have the optimum visibility with the least trouble! We have already discussed how you need to post regularly to gain more attention, but posting too often can also be a problem – you will end up being that irritating account that keeps popping up on your followers' feeds with promotional mumbo-jumbo! Do that and it is very likely that you will be unfollowed.

Getting the timing right can be a tricky thing – Schedugram takes away that headache so that you can focus on what you post instead of when to post. Social media managers across the globe who manage big brands and big companies make use of Schedugram to get more visibility for their clients; it is the best way to manage your Instagram account without too much hassle.

Here are some of the benefits of Schedugram:

- Post regularly on a schedule optimized for your brand. Schedugram does not work out a strategy   for you to post; you must do yourself. What it does do is *schedule* your posting. For example, you can put up an image today and schedule for it to be posted on Friday of the next week at exactly 6 pm. Schedugram will do the posting for you; you do not even need to log into your account! All you have to do is to enter the preferred timing and date you want your content to go up on and it is done automatically. While you still have to plan and strategize the best times to post, this gives you more leeway in terms of worrying about the content instead of the actual posting itself. You can focus on getting your picture, caption, and tags just right and then give it to Schedugram, which will then post it as per your specifications. This tool is especially useful when you are planning a long campaign, as it lets you keep track of your posts and product to make sure you reach your customers exactly on time.

- Managing a single account on Instagram itself is confusing and hectic; multiple account holders must be on their toes and ready to constantly engage their followers! One problem with Instagram is you need to keep logging out of one account and then log in to the other to gain access to it. Schedugram takes away that issue and lets you manage multiple accounts simultaneously. A lot of big brands have more than one account to manage; Schedugram saves them time when it comes to switching between many accounts. They can schedule their different posts for each account on a single device without repeated logins. When using Schedugram, all you need to do is toggle the right account and schedule and make sure you have put up the details correctly. Then, Schedugram will do your posting on that account for you!

- Schedugram also allows you to share your Instagram account with others and make it a multiple user account. This is especially useful for business owners, as you can now allow your employees and trusted workers access to your brand"s Instagram account. You can get your team members to schedule posts and upload drafts and pictures. You can also check to see who has scheduled what so that it all still remains under your explicit control, giving you the accountability you require to run your business. In essence, this makes it sort of a community platform, which is an excellent way of building trust and loyalty within the company itself, since your employees will feel good about the access and the trust you are giving them.

- You can create content on Schedugram itself; it offers an image editor that is extremely easy to use. You can edit your pictures and images and then upload them instantly. There are features to let you crop your picture, add interesting filters to them, add text, fix colors and shades, rotate them to different angles, and many more options. But it does not stop there! Schedugram provides you with the option to directly integrate with *Canva*. For those who do not know, *Canva* is one of the best graphic design apps out there that helps you make the best pictures you can. It is a photo editing software of extremely good quality, and Schedugram lets you access to your *Canva* account so your images to be uploaded to Instagram are of the best possible quality.

- You can also post images in bulk instead of uploading them one by one. If you do not need to do much in the way of editing your images, just use the bulk upload tool to put up a large number of images, add your captions, and pick out dates and times to get them all posted without hassle. You can also easily save your pictures in draft mode, edit them, and then post at a later time.

- You pay for each of the accounts that you add to your Schedugram base, but you can remove any of those at any given time. Payment plans happen on a pro-rated basis, so you can manage accounts without difficulty. And if you add more than a specified number of accounts, Schedugram gives you a good discount on your payment options. The best part is that if you do not like a particular tool or if you need something customized to suit your specific business needs, Schedugram is willing to take the time to sit with you and work it out, so you get the best possible service.

- You can use the seven-day trial to check out the app and see if it will be useful to you. The trial is, obviously, free and you get almost all the benefits so you will know if it will suit your business and your account. Once your trial ends, you can pay your account fee and start using Schedugram in full!

Schedugram is extremely cheap; it costs about 20 dollars a month only for a single Instagram account and international taxes are not levied. This means that you can sit in any part of the world and access your account at the same cost as everyone else! You can also pay your fees at an annual rate and receive a discount equivalent to fees charged for two months. Payment methods, though, are specific to the type of accounts – credit cards are accepted for all, but things like PayPal are applicable to select accounts, so make sure you check out your options before you sign up. You can also cancel your subscription to Schedugram at any time, as long as you do so before your next billing cycle. If you are on the annual plan, of course, this is a bit of a hassle. Get in touch with customer support and they will help you out.

One thing you do need to keep in mind is that Schedugram is not endorsed or affiliated with Instagram or Facebook in any manner. They are a third party company who has been set up to help small business owners manage their marketing strategies on a continually evolving, multifaceted social media platform. This means you will need to give them your Instagram login details, like your username and password, and they will store it for their use.

As you can guess, the biggest problem most people have with this is the issue of safety. How can you trust a third-party commercial website with such sensitive information, especially as a proprietary business venture? Schedugram does assure their client of safe storage; they take client confidentiality very seriously and your details are not shared with anyone under any circumstance. They have employed a number of security measures to protect their clients, which you can read about on their website, and they do reassure their users it will all be safe and secure.

They also offer you another option: a demo during your free trial. This is a very good option to not only check out their security, but to also see how well the application and software work for you and your business. You can create a dummy account on Instagram to run yourself a quick test – sign up on Schedugram and submit your temporary account details. Then, start using the various features to see how it works for you. You will be able to gauge both its effectiveness as well as its security measures!

Schedugram also lets you use a version of their software that you can put up on your host network instead of the public domain. This means you are in control of your details and you have full access to see who has your account information and who can make use of this information. Obviously, this is the best option for any business venture, so get in touch with the customer support team and find out more details about how to set it up!

As you can see, Schedugram is a brilliant tool you can use to manage your Instagram posts. It is simple, easy to use, cheap, and effective – just sign up, try out the free trial version, and then make use of it to gain more visibility on Instagram!

Iconosquare

There is no use running an Instagram account if you cannot check how much of an effect it is having on your business and your overall marketing strategy. As we have said before, advertising on Instagram must only be a part of a larger plan – it cannot be the end all and be all of your entire marketing plan. But even on Instagram, you need to analyze how well you are doing and how much traffic is coming to your account, as well as how many customers are becoming part of your loyal followers. Without this type of analytics, you will not be able to improve or change your current marketing outline.

Iconosquare is an excellent tool that lets you complete all these analytics. It gives you information and insights into your account and the follower traffic so you will be able to break down your work into smaller segments that you   can further analyze. It gives you the key metrics about your Instagram account so you can see where you are going wrong, what you need to correct, or even what you are getting right and how you can further augment that particular area.

The software gives you details like the number of „likes" you get for a picture, what your most liked pictures are, what the average number of likes you can expect for a picture, the average number of comments you usually get, the growth rate of your followers, etc. But it does not stop at providing you with advanced analytical data alone; it allows you other added benefits like promoting your account via other platforms such as Facebook and Twitter, managing contests, engaging with your customers, and more.

Here are some of the key features that Iconosquare promises:

1. *Analytics* – As I have said, Iconosquare is one of the prime Instagram analytics tools that can help you analyze the facts and figures when it comes to your business accounts. In their own words, they cover the "most important aspects of your Instagram presence." Complete with charts and graphs and the like, they

provide you with all the details you require to see how well your account is doing.

Activity-wise, they keep an eye on all that you do on Instagram and how it affects your business. For instance, they will give you statistics based on how many people have followed or unfollowed you within a time period, what the average growth rate is of your follower base, the follower response to your posts, etc. With that data in hand, they go on to analyze the details such as when is the best time to post so you can get maximum engagement with your followers, how you can further engage them, and more.

Iconosquare will also analyze your relationship with your followers. They will give you numbers such as how many likes and comments you are getting, how many you are liable to get on an average, how your likes and comments section have grown or shrunk, what the reasons (i.e., what particular posts) for this are, the rates at which your account is gaining visibility and being mentioned on Instagram, etc. Other than you, Iconosquare also takes into account your followers — who are the followers you engage the most with? Are they well followed, thereby giving you more visibility via their accounts? Who are your new followers and what are the statistics related to them and their accounts? Who are the followers who unfollowed you and how much of a difference is the loss of their account going to be to your overall Instagram presence? Who are the accounts you should follow back and build a reciprocal relationship? Who are the accounts you can afford to not follow? All these follower account details will give you a clearer picture of what your standing in Instagram is at the moment.

Iconosquare also builds reports, charts and graphs. All the key metric data pertaining to a month is categorized, sorted, and then put into detailed visual graphics for you to peruse and understand. The reports will tell you everything you need to know at a glance. What is the daily growth or loss rate of your followers? Demographically speaking, who are your top followers, where are they located in the world, and what is their particular reach within their community? What are your top posts? How can they be further sorted into different categories based on the number of likes and comments and how well do they engage your audience? What is the best time to post? Which hashtag has given you the best performance and which has given you the worst? Which filters work best for you and which are not all that great?

All these details are sorted out into neat little charts, diagrams, and reports you can quickly skim for an overview. Detailed statistics follow visuals, so you can easily analyze your account and see how and where you need to change or improve.

Taking off from the last two questions in the previous paragraph, tracking hashtag and filter performances is one of the most important Instagram analytics you need. As we have seen, hashtags are your primary marketing tool. The more relevant and pertinent your tag, the better your visibility. Iconosquare expressly offers you the track your hashtag performance option, which will allow you to see where you are doing it wrong (or right) in hashtagging.

You can measure the growth of your tags by analyzing the number of posts that are put up using that particular tag. Obviously, the comments and the likes that both your posts and the other posts get with the tags you have created are a clear indication of how popular your tag is becoming. The more popularity your post gains, the more visibility you get. Iconosquare also helps you identify the top Instagrammers who use your hashtag over a specific period. These accounts will serve as your brand advocates. Statistics about them, such as where they live and how much visibility they have as an Instagrammer should be taken into account when planning marketing strategy.

Iconosquare also tracks who is posting using your hashtag. What did you have in mind when you created it? What are people actually posting with it? Are the two related or has it taken a life of its own? Is it good or bad for your company? Take all this data into account when outlining your further marketing plan.

Iconosquare also allows you to check out your competitors" analytics. Any business must be up to date with the changes and inventions their competitors put up; this will allow you to see how well they perform on Instagram. Iconosquare tracks any account you show an interest in – it tracks your top five competitors and analyzes what the rate of their Instagram growth is. How many posts do they make in a month? What is their follower and engagement rate? How well does the media that they post perform? How many likes, comments, and shares do they get and how well do they engage their audience? What is the average number of comments and likes they are likely to get and how does that relate to your own performance? Thus, when you know how well your competitors are doing, you know your own standing in the business and what you need to do make it better.

2. *Multifaceted Website View* – Instagram is a mobile app, which means you are limited to a few screen options on a small device. Even if you do access it from your laptop, you really don't have the option of posting across platforms or browsing with other media networks and the like. Iconosquare gives you another viewing option that is multifaceted.

You have all your usual Instagram options, such as your Instagram feed, your likes, your space to post comments and pictures, following and unfollowing accounts, etc. But now, you can also have a record of all the people you have already liked or commented on, all the photos and videos of the people you follow, the details of popular posts, etc., but in one screen. You also have your own user profile on the same screen, so you can access everything at once and then see where you need to make a change.

Key features are the ability to save and back up your Instagram photos easily, share on other social media platforms like Pinterest, Facebook, Tumblr and Twitter (they link you to another seven social media platforms, so make full use of this option!), reposting the photos you enjoy and like without any hassle.

It doesn't seem like much, but this kind of an improved view and user interface makes a difference when you are in a hurry and pressed for time. You get easy access to everything you need in a matter of seconds. You no longer have to scroll through mounds of data to find what you're looking for!

3. *Manage Conversations and Engagement with Your Audience* – Given that Instagram is an open platform for your followers to comment on your posts and interact with each other while still interacting with you too, the comments section and conversation threads can end up super messy. It can be very difficult to scroll through and then properly respond to all those comments that need replies. You don't want to lose a customer who has asked a question but received no reply because your account became so cluttered!

Iconosquare lets you overcome this issue by allowing you to manage your conversations and track your comments. They help you sort your account out so you do not ever miss a comment from a stakeholder or find it too hard to respond to them. You will receive notifications as soon as new comments are posted on any of your media. Each comment is tracked and recorded so you know whom to respond to at what time. You also have the option of replying to all comments or replying to each one individually, even with your favorite emojis!

With Iconosquare, you can also mark the comments you receive as read or unread, the latter option being for those who wish to come back to them later. You can also delete comments you find inappropriate, thereby keeping your feed clean and healthy.

This way, no customer or follower feels ignored, despite the large number of people with whom you interact. Engagement becomes healthy and grows at a good rate!

4. *Promotional Techniques* – Iconosquare offers you an excellent option to cumulate all your marketing strategies across different social media platforms. They help you increase your exposure and visibility across the entire Internet by linking you to your different accounts; for instance, even your Facebook fans who do not follow you or have an Instagram account can check out your content with the Instagram feed tab for Facebook. You can also get your current tags promoted and premium

members get up to five hashtag feeds with comprehensive analytics on their performances across the different media platforms.

Iconosquare lets you play with the Instagram and Facebook crossover – if you want to, you can get them to design your Facebook Timeline Cover using your Instagram posts. This gets your Instagram account more visibility across Facebook *and* gives you a professional looking timeline that attracts attention!

The photo gallery widget lets you play with the different pictures you have that you can post across social media. Using this, you can share your own hashtags and content, play with the layout that will match your different pages, and then post wherever you think you will get more viewers! Iconosquare also offers you the option of a *vanity URL* – this will activate a public page for you and your account. This way, even those who do not have an Instagram or even a Facebook account will be able to look through your profile and gain access to your content.

Obviously, the applications of these are limited to being part of an overall strategy, but they are excellent methods to augment your marketing!

5. *Running and Managing Contests* – We have already looked in detail at what type of contests you can host and how you should go about this. Contests are not easy to keep track of; with the huge number of people taking part, you need to be on your toes and make sure every customer is catered to. Iconosquare helps in managing contests – in fact, they offer a step-by-step guide on how to do it!

- Step 1: launching your contest. Iconosquare is pretty much willing to do all the work for you, so all you need to do is schedule your contest starting, ending, and posting dates, then put them up. Key in the hashtags you want associated with the contest and Iconosquare will take care of the rest.

- Step 2: manage the entries you receive. Iconosquare lets you display all the entries wherever you want them to be displayed; a landing page for your contest is also created and hosted on Iconosquare. They also allow you the option of installing a tab on your Facebook page that is customized. This will let you feature your contest and then interact with your audience.

- Step 3: Iconosquare determines how you can increase your audience" s participation by cutting across different social media platforms. One reason why an Instagram contest tends to be restricted to Instagram is how tough it can get to track entries from different social media, not just on Instagram! Now, Iconosquare helps you track it all, so you can invite entries from people with Facebook, Twitter, and other accounts too! The participants can also do a direct upload from any device, giving them access to your contest easily.

- Step 4: moderate entries to ensure nothing untoward is posted. Bulk moderation is also offered as an option. You can approve, preselect, or even refuse entries enmass instead of going through them all one by one, thereby saving you precious time and effort. You also receive notifications each time new media is added, and the likes and comments are analyzed and moderated according to your preferences.

- Step 5 : identifying how successful and effective your contest has been in its purpose.

There is no point running a contest if you do not analyze its performance. Iconosquare helps you do this by giving you charts and insights into your contests. How much visibility you gained, how many people participated, how much of your target audience was engaged, etc. – all of these questions are answered. The data is also available for you to export into reports of your own so that you can sort it out according to your specific preferences.

These are just some of the major features Iconosquare provides. It is an excellent analytical tool to check your performance, so sign up and make full use of it to see how well your Instagram marketing strategy is working!

Websta

Earlier known as Webstagram and rechristened Websta, this is one of the most popular analytical tools for Instagram users. It helps you find the best hashtags for your posts and gives you statistics on the latest trends to help you perform even better. Initially begun as a web viewer for Instagram, it has since then expanded to include value added services like the Websta Board, Instagram Gallery Widget, and the like.

Here are a few of its key features:

- Websta Board – You can use this to organize your Instagram posts into proper groups and manage the content you posted online.

- Search – You can search for the most popular hashtags and get statistics on them so you can incorporate them into your own posts. There is also the „user keywords" option; these are like hashtags for specific accounts.

- Instagram Gallery Widget – You can turn any feed of yours into a gallery, be it a hashtag feed, a location feed, or even your own profile feed! You can then get this displayed on your blog or your website.

- Follow Me Widget – This allows any visitors to your website or your blog to access your Instagram content without directly using Instagram.

- Repost – You can repost any picture or post of a user with this option; it even adds the watermark of the user" s name so your content is not flagged as stealing or someone else' s work.

- Share – Again, you can easily share posts on Websta as well as other social media platforms.

- My Stats – As the name suggests, this will let you know the facts and figures about your Instagram account. You can check how well your followers and other users are engaging with your posts and change or modify your strategy to get optimum interaction and visibility.

- Visitors – Again, this offers you stats on who is checking out your website and what the effect of such traffic is having on your account.

- Popular Posts – Websta collates all the most popular hashtags and posts so that you can analyze and peruse them for ideas, changes, and any needs of your own.

Ultimately, Websta is a way of accessing and using Instagram from your computer. Use it well to augment your strategy, as it gives you the top posts, hashtags, and users. Check them all out to see what's trending and what is new so your own account is fresh and engaging!

These are just a few of the many tools available online to make sure your Instagram account is well marketed and receives good traffic. Check them all out. Remember, each business is different, so you will need to pick the tools that best suit your needs. Most of them offer a free trial or a demo version, so go in for that, check out the features, and then decide which one works for you!

Conclusion

Now that you have come to the end of this book, we would first like to express our gratitude for choosing this particular manual. We have tried to accumulate as much information as possible in order for you to be able to become Instagram experts.

If you already have an Instagram profile, you can now use it in a much more efficient and maybe even profitable manner. If you don't have one yet, you can get started right away. Use the material here to begin and work your way through this service to the top-most popular users they have.

The information and pointers in this book will help you create an Instagram profile that has the potential to become amazing over a very short period of time. Use these to be a successful Instagram entity and recommend it to your friends and family as well.